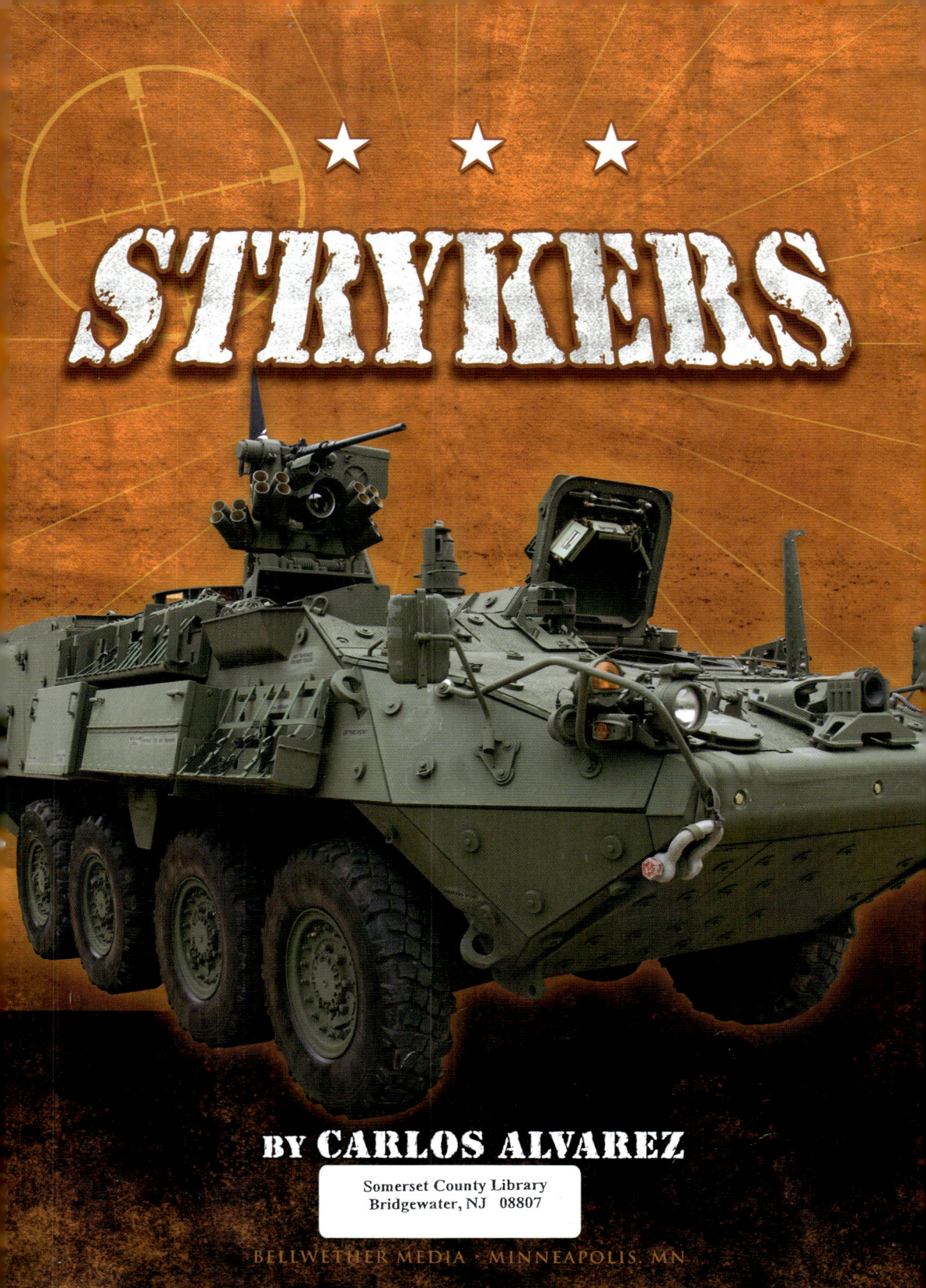

Are you ready to take it to the extreme? Torque books thrust you into the action-packed world of sports, vehicles, and adventure. These books may include dirt, smoke, fire, and dangerous stunts. WARNING: read at your own risk.

Library of Congress Cataloging-in-Publication Data

Alvarez, Carlos, 1968-
  Strykers / by Carlos Alvarez.
      p. cm. – (Torque: Military machines)
  Summary: "Amazing photography accompanies engaging information about Strykers. The combination of high-interest subject matter and light text is intended for students in grades 3 through 7"–Provided by publisher.
  Includes bibliographical references and index.
  ISBN 978-1-60014-496-7 (hardcover : alk. paper)
  1. Stryker armored vehicle–Juvenile literature. I. Title.
  UG446.5.A425 2010
  623.7'475–dc22
                                                2010000863

This edition first published in 2011 by Bellwether Media, Inc.

No part of this publication may be reproduced in whole or in part without written permission of the publisher. For information regarding permission, write to Bellwether Media, Inc., Attention: Permissions Department, 5357 Penn Avenue South, Minneapolis, MN 55419.

Text copyright © 2011 by Bellwether Media, Inc. TORQUE and associated logos are trademarks and/or registered trademarks of Bellwether Media, Inc.

The images in this book are reproduced through the courtesy of: Chung Sung-Jun/Getty Images, pp. 4-5, 9 (top), 19; Ted Carlson/Fotodynamics, pp. 8-9, 18; Getty Images, pp. 20-21; all other photos courtesy of the United States Department of Defense.

Printed in the United States of America, North Mankato, MN.
010111   1183

CONTENTS

THE STRYKER IN ACTION	4
ARMORED COMBAT VEHICLE	10
WEAPONS AND FEATURES	14
STRYKER MISSIONS	18
GLOSSARY	22
TO LEARN MORE	23
INDEX	24

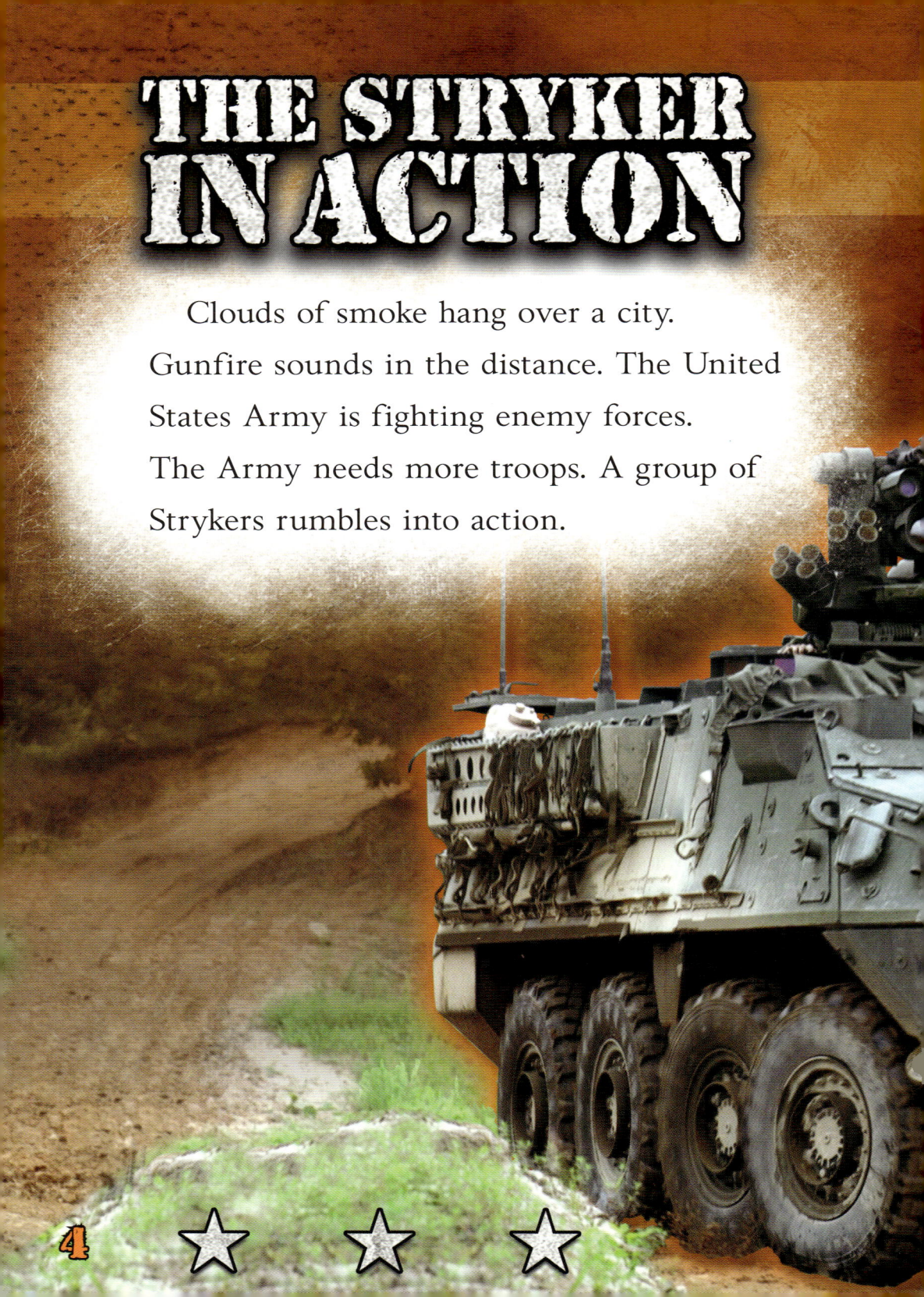

THE STRYKER IN ACTION

Clouds of smoke hang over a city. Gunfire sounds in the distance. The United States Army is fighting enemy forces. The Army needs more troops. A group of Strykers rumbles into action.

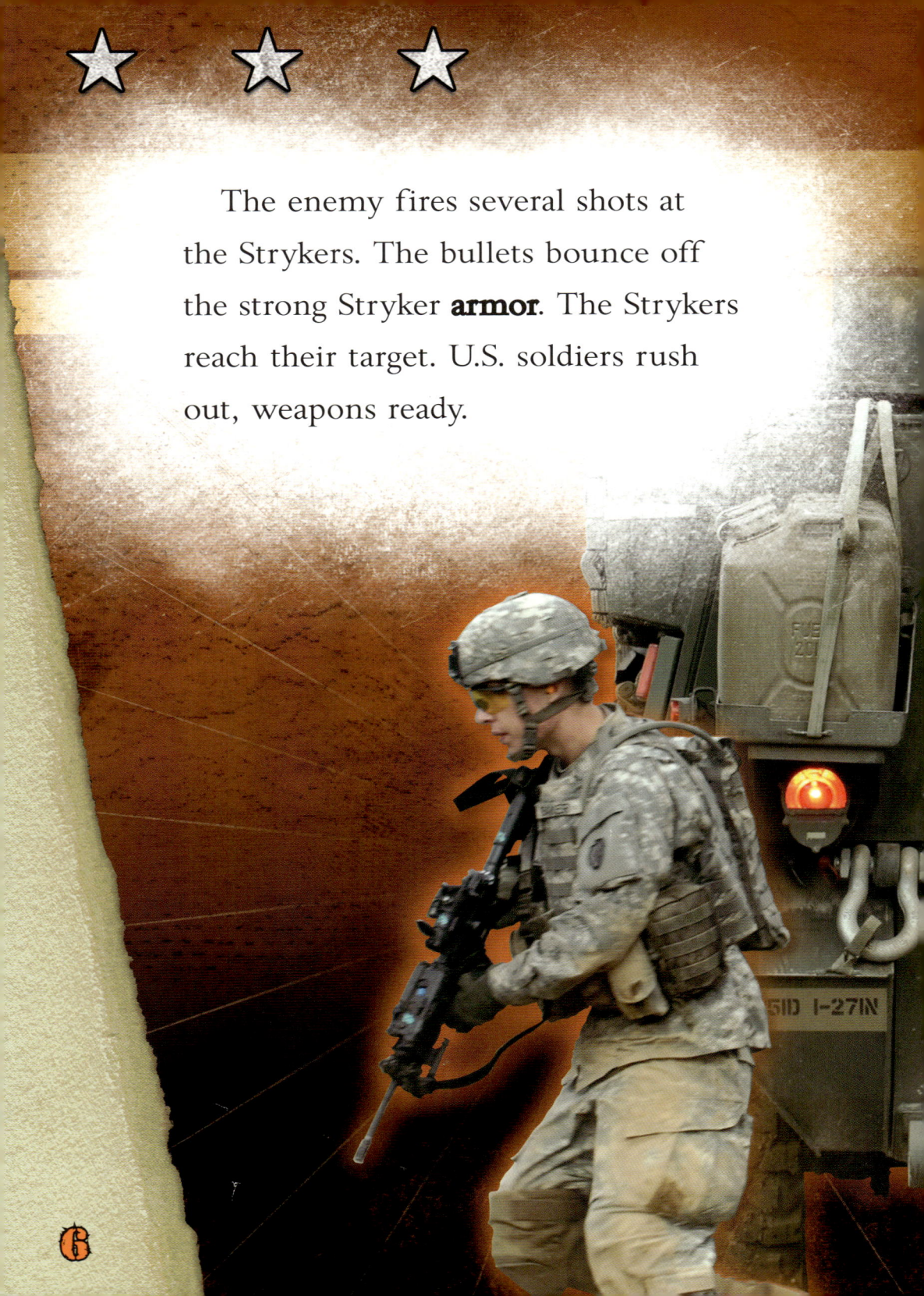

The enemy fires several shots at the Strykers. The bullets bounce off the strong Stryker **armor**. The Strykers reach their target. U.S. soldiers rush out, weapons ready.

FAST FACT

The United States Army has about 3,000 Strykers in service.

The new soldiers join the battle. The Strykers stay and support them. They fire their guns at the enemy. They won't leave until the enemy is defeated.

ARMORED COMBAT VEHICLE

The Stryker is an eight-wheeled armored combat vehicle. The U.S. Army has two main types of Strykers. The Mobile Gun System (MGS) carries a large cannon. The Infantry Carrier Vehicle (ICV) carries up to nine soldiers. There are other Stryker **variants**. They are less common than the MGS and ICV.

M1128 Mobile Gun System (MGS)

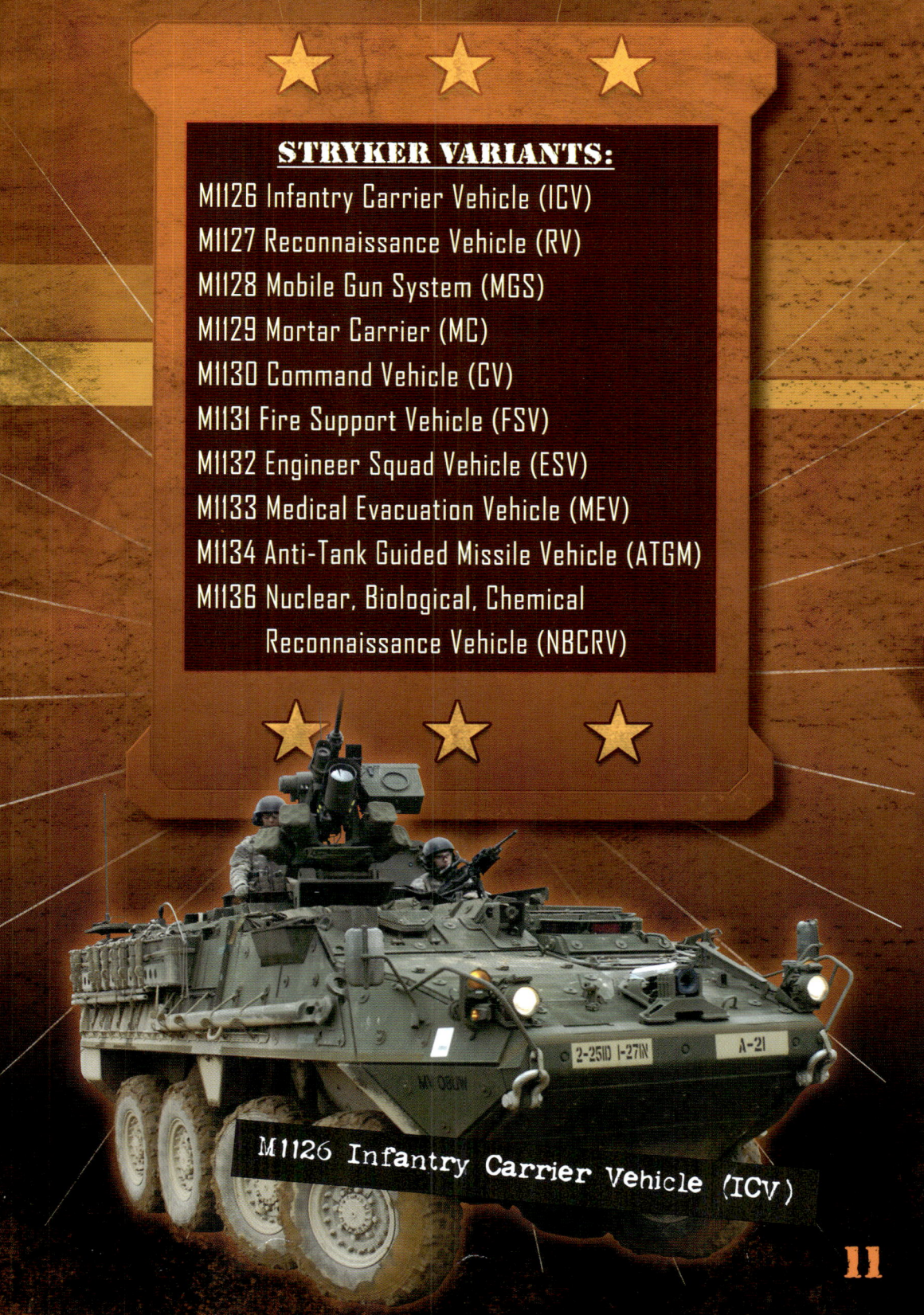

STRYKER VARIANTS:

M1126 Infantry Carrier Vehicle (ICV)
M1127 Reconnaissance Vehicle (RV)
M1128 Mobile Gun System (MGS)
M1129 Mortar Carrier (MC)
M1130 Command Vehicle (CV)
M1131 Fire Support Vehicle (FSV)
M1132 Engineer Squad Vehicle (ESV)
M1133 Medical Evacuation Vehicle (MEV)
M1134 Anti-Tank Guided Missile Vehicle (ATGM)
M1136 Nuclear, Biological, Chemical Reconnaissance Vehicle (NBCRV)

M1126 Infantry Carrier Vehicle (ICV)

The Stryker has the power of a tank and the control of a **Humvee**. It can go where larger armored vehicles cannot go. It is most useful in crowded areas such as cities.

The Stryker entered Army service in 2002. It was named after Robert F. Stryker and Stuart S. Stryker. Both men were soldiers killed in action. They were both awarded the **Medal of Honor**.

★ FAST FACT ★

Strykers have quieter engines than other troop carriers. This often allows them to move without being heard by the enemy.

WEAPONS AND FEATURES

The Stryker has features that help it in battle. It is covered in strong armor. The armor includes layers of steel and **ceramics**. It stops most kinds of gunfire. Some Strykers have extra armor. Cage armor can protect the Stryker from **rocket-propelled grenades (RPGs)**.

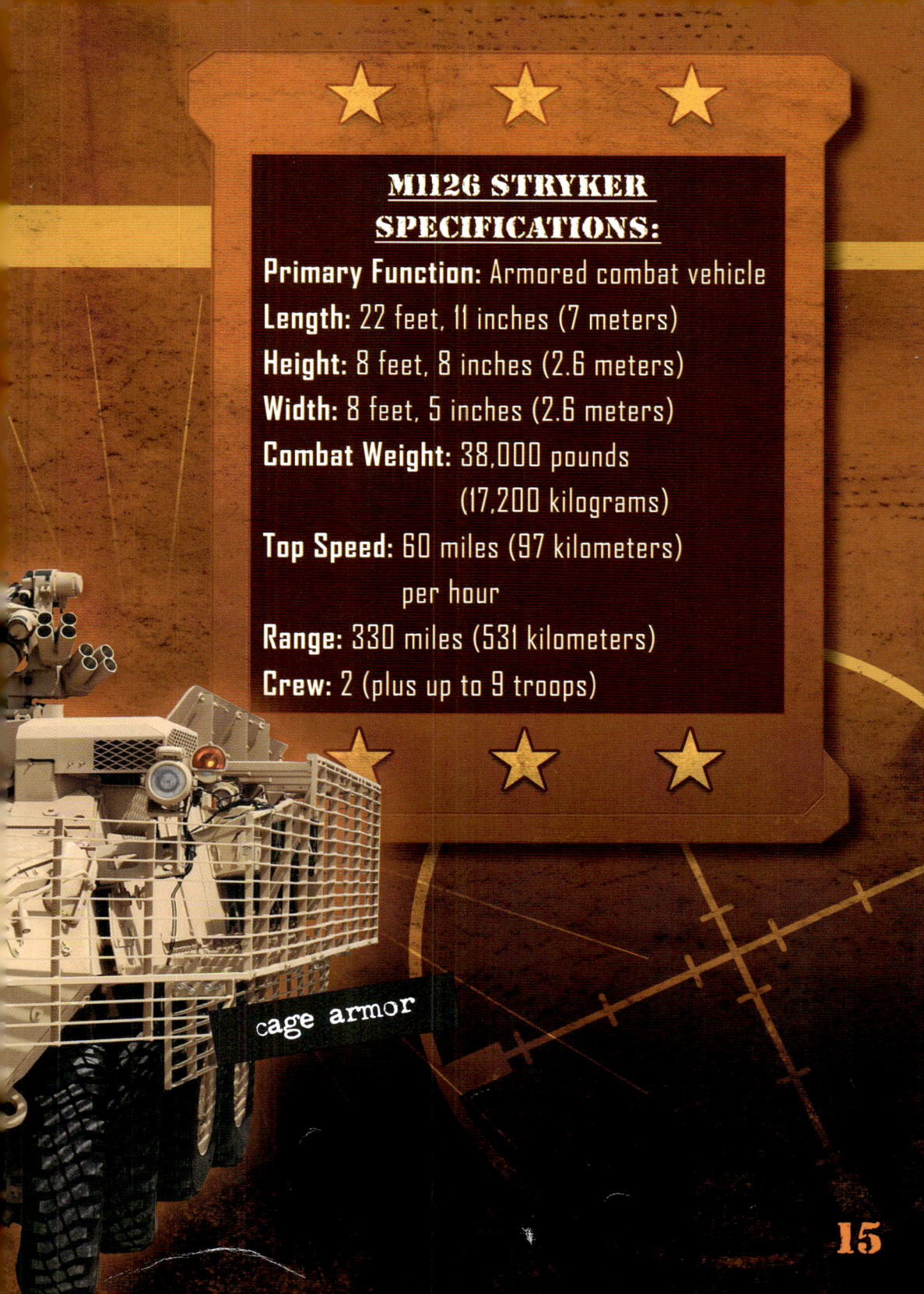

M1126 STRYKER SPECIFICATIONS:

Primary Function: Armored combat vehicle
Length: 22 feet, 11 inches (7 meters)
Height: 8 feet, 8 inches (2.6 meters)
Width: 8 feet, 5 inches (2.6 meters)
Combat Weight: 38,000 pounds (17,200 kilograms)
Top Speed: 60 miles (97 kilometers) per hour
Range: 330 miles (531 kilometers)
Crew: 2 (plus up to 9 troops)

cage armor

Different variants of the Stryker have different weapons. The ICV unit has a .50-caliber M2 machine gun. It also has either a 7.62mm machine gun or an MK19 **grenade launcher**. The MGS unit's main gun is a **turret**-mounted 105mm cannon. The MGS also has an M2 machine gun. Other variants carry **missiles**, **mortars**, and more. All Stryker variants have M6 smoke grenade launchers. Smoke grenades let off huge amounts of smoke. All of these weapons help Strykers complete their **missions**.

missile launcher

STRYKER MISSIONS

Strykers can perform a wide range of combat missions. They most often carry soldiers. Some variants serve as scout vehicles, ambulances, or command centers.

FAST FACT

Strykers can be carried in C-130 Hercules transport planes.

Two crew members work together to operate a Stryker. The driver steers the vehicle. The **commander** controls the Stryker's sensors and weapons. Both the driver and the commander can operate weapons.

Strykers offer a great combination of power and speed. The many variants can accomplish many different missions. This ensures that the Stryker will remain a big part of the U.S. Army for years to come.

GLOSSARY

armor—protective plating

ceramics—hard materials made by heating and cooling materials such as clay

commander—the crew member in charge of a Stryker

grenade launcher—a weapon that fires small explosives called grenades

Humvee—a multipurpose armored military vehicle

Medal of Honor—the highest military honor awarded by the United States government

missiles—explosives launched at targets on the ground or in the air

missions—military tasks

mortars—short-barreled guns that launch explosives high into the air

rocket-propelled grenades (RPGs)—small explosive rockets that can be fired from a handheld launcher

turret—a weapon mount that can rotate in any direction

variants—versions of something; there are many variants of the Stryker that do different jobs.

TO LEARN MORE

AT THE LIBRARY
Baker, David. *M1126 Stryker.* Vero Beach, Fla.: Rourke Publishing, 2007.

David, Jack. *United States Army.* Minneapolis, Minn.: Bellwether Media, 2008.

Hama, Larry, and Bill Cain. *Tank of Tomorrow: Stryker.* New York, N.Y.: Children's Press, 2007.

ON THE WEB
Learning more about military machines is as easy as 1, 2, 3.

1. Go to www.factsurfer.com.

2. Enter "military machines" into the search box.

3. Click the "Surf" button and you will see a list of related Web sites.

With factsurfer.com, finding more information is just a click away.

INDEX

2002, 13
.50-caliber M2 machine gun, 16, 17
7.62mm machine gun, 16
105mm cannon, 16, 17
ambulances, 18
armor, 6, 14, 15
C-130 Hercules, 19
ceramics, 14
command centers, 18
commander, 20
crew members, 20
driver, 20
engine, 13
Humvee, 12
Infantry Carrier Vehicle (ICV), 10, 11, 16
M6 smoke grenade launcher, 16
Medal of Honor, 13
missiles, 16
missions, 16, 18, 21
MK19 grenade launcher, 16
Mobile Gun System (MGS), 10, 16
mortars, 16
rocket-propelled grenades (RPGs), 14
scout vehicles, 18
soldiers, 6, 8, 10, 13, 18
steel, 14
Stryker, Robert F., 13
Stryker, Stuart S., 13
turret, 16
United States Army, 4, 7, 10, 13, 21
variants, 10, 11, 16, 18, 21

J 623.7475 ALV
Alvarez, Carlos, 1968-
Strykers

WARREN TWP LIBRARY
42 MOUNTAIN BLVD

JAN 3 1 2012

WARREN, NJ 07059
908-754-5554